LIST OF BIRDS COLLECTED IN THE ISLAND OF BOURU

(ONE OF THE MOLUCCAS), WITH

DESCRIPTIONS OF THE NEW SPECIES

(1863)

BY

ALFRED RUSSEL WALLACE

British Library Cataloguing-in-Publication Data
A catalogue record for this book is available from the
British Library

Alfred Russel Wallace

Alfred Russel Wallace was born on 8th January 1823 in the village of Llanbadoc, in Monmouthshire, Wales.

At the age of five, Wallace's family moved to Hertford where he later enrolled at Hertford Grammar School. He was educated there until financial difficulties forced his family to withdraw him in 1836. He then boarded with his older brother John before becoming an apprentice to his eldest brother, William, a surveyor. He worked for William for six years until the business declined due to difficult economic conditions.

After a brief period of unemployment, he was hired as a master at the Collegiate School in Leicester to teach drawing, map-making, and surveying. During this time he met the entomologist Henry Bates who inspired Wallace to begin collecting insects. He and bates continued exchanging letters after Wallace left teaching to pursue his surveying career. They corresponded on prominent works of the time such as Charles Darwin's *The Voyage of the Beagle* (1839) and Robert Chamber's *Vestiges of the Natural History of Creation* (1844).

Wallace was inspired by the travelling naturalists of the day and decided to begin his exploration career collecting specimens in the Amazon rainforest. He explored the Rio

Negra for four years, making notes on the peoples and languages he encountered as well as the geography, flora, and fauna. On his return voyage his ship, Helen, caught fire and he and the crew were stranded for ten days before being picked up by the Jordeson, a brig travelling from Cuba to London. All of his specimens aboard Helen had been lost.

After a brief stay in England he embarked on a journey to the Malay Archipelago (now Singapore, Malaysia, and Indonesia). During this eight year period he collected more than 126,000 specimens, several thousand of which represented new species to science. While travelling, Wallace refined his thoughts about evolution and in 1858 he outlined his theory of natural selection in an article he sent to Charles Darwin. This was published in the same year along with Darwin's own theory. Wallace eventually published an account of his travels *The Malay Archipelago* in 1869, and it became one of the most popular books of scientific exploration in the 19th century.

Upon his return to England, in 1862, Wallace became a staunch defender of Darwin's landmark work *On the Origin of Species* (1859). He wrote responses to those critical of the theory of natural selection, including 'Remarks on the Rev. S. Haughton's Paper on the Bee's Cell, And on the Origin of Species' (1863) and 'Creation by Law' (1867). The former of these was particularly pleasing to Darwin. Wallace also published important papers such as "The Origin of Human

Races and the Antiquity of Man Deduced from the Theory of 'Natural Selection" (1864) and books, including the much cited *Darwinism* (1889).

Wallace made a huge contribution to the natural sciences and he will continue to be remembered as one of the key figures in the development of evolutionary theory.

Wallace died on 7th November 1913 at the age of 90. He is buried in a small cemetery at Broadstone, Dorset, England.

LIST OF BIRDS COLLECTED IN THE ISLAND OF BOURU

(ONE OF THE MOLUCCAS), WITH DESCRIPTIONS OF THE NEW SPECIES

(1863)

This collection of birds was made by myself during two months of the year 1861. It consists of *sixty-six* species, among which were no less than *seventeen* new ones. Of these, *three* were found about the same time in the Island of Sula, and, with a new *Pitta*, have already been described in the Society's 'Proceedings,' leaving *thirteen* to be described in the present paper.

In my paper "On the Birds of the Sula Islands," read before the Society at their last Meeting, I pointed out that the large proportion of purely Celebes forms found there forced us to the conclusion that a closer connexion had once existed between those islands and Celebes, and required us to class them as forming a single zoological group. The Island of Bouru must, on the contrary, be classed with the Moluccas; for, leaving out about *twenty* species of rather wide distribution, the remaining *forty-six* are all either identical with, or most nearly allied to, Moluccan species.

Not a single characteristic Celebes form is found in Bouru; and there are only *three* birds in the island whose affinities seem rather with the Indian than the Australian region, viz. *Alcedo moluccensis*, *Hirundo javanica*, and *Treron aromatica*.

Bouru is therefore the western limit of the Moluccan fauna, and is the poorest portion of it, having several very remarkable deficiencies. *Lorius*, found in every other island of the Moluccas and New Guinea, is absent; *Cacatua*, found in every island of the Australian region, is also absent; and, stranger still, *Buceros* and *Corvus*, found in almost every large island of the archipelago, are both wanting. With these exceptions, most of the Moluccan types are represented either by identical or allied species.

The following is a list of the new species now described, and of a few others which seem confined to Bouru:--

Tanygnathus affinis, n.s., Bouru and Ceram.
Accipiter rubricollis, n.s., Bouru and Ceram.
Athene hantu, n.s.
Tanysiptera acis, n.s.
Ceyx cajeli, n.s.
Pitta rubrinucha, Wallace.
Cisticola rustica, n.s.
Mimeta bouruensis (H. & J.).
Criniger mysticalis, n.s.
Monarcha loricata, n.s.

Rhipidura bouruensis, n.s.
Tropidorhynchus bouruensis, n.s.
Campephaga marginata, n.s.
Dicæum erythrothorax, Less.
Nectarinea proserpina, n.s.
Gallinula frontata, n.s.

All but the two first species in this list are confined to Bouru only, and they are mostly representative species of Moluccan forms. Besides these, the three species of *Pachycephala* are also, as far as the Moluccas are concerned, peculiar to Bouru; for though they are found also in Sula, they have evidently emigrated there, the Celebes group, to which Sula belongs, not possessing any species of the genus. The Island of Bouru may therefore be considered to have added *seventeen* new species, but not any new forms or genera, to the Moluccan avifauna.

GEOFFROIUS PERSONATUS.

Psittacus personatus, Shaw.

P. bataviensis, Wagl. Mon. Psitt. p. 624.

Hab. Bouru, Amboyna, Ceram, Goram, Ké and Aru Islands.

Remarks.--The specimens from Bouru, and some from Ceram, are 12 1/2 inches long; that from the Aru Islands 9 inches; but I have a series of intermediate sizes, and can discover no differences of form or in the distribution of the colours. I must therefore consider Mr. G. R. Gray's *Psittacus aruensis* (P. Z. S. 1858, p. 183) as only a small variety of this species, and his *P. capistratus*, from the Ké Islands (*ibid.* p. 183), as a young male bird of the same species.

ECLECTUS MAGNUS.

Psittacus magnus, Gm. S. N. i. p. 344.

Hab. Bouru and the other islands of the Moluccas and New Guinea.

ECLECTUS PUNICEUS.

Psittacus puniceus, Gm. ? (et auct.) Pl. Enl. 518.

Hab. Bouru, Amboyna, and Ceram.

Remarks.--This bird is sufficiently distinct from the *Psittacus grandis*, Gm., which is confined to the Gilolo group, in its smaller size, duller red colour, red under tail-coverts, and tail only orange-tipped, in place of the yellow under tail-coverts and larger yellow tail-band of *E. grandis.* Great confusion exists in the synonymy of the *Psittaci*, owing, I believe, to the fact of so many of these birds having been described from specimens which have lived a long time in confinement, and have acquired abnormal colours in various parts of their plumage. The production of such coloured variations is, in fact, an art practised by the native tribes both in South America and in the Eastern Islands. Another cause of error is from young birds having been described; and a third, from the deficiencies of badly prepared native skins having been made up by the addition of parts (often the wings and tail) of other birds. In the present case I have little doubt that this bird is the *P. puniceus* of Gmelin, and the *Lorius amboinensis* of Brisson, whose description, generally so eminently accurate, appears to apply to a young bird which had lost its primary quills. I cannot agree to the revolution in nomenclature proposed by Mr. G. R. Gray, in using the names of Boddaert, which have been considered of no authority by every other author from the time of Gmelin to that of Prince Bonaparte.

TANYGNATHUS AFFINIS.

Viridis, subtus flavescens; capite saturate viridi; dorsi plumis cæruleo marginatis; crisso cæruleo; tectricibus alarum minoribus et mediis obscure viridibus, flavo marginatis, versus marginem et flexuram alarum viridi-cæruleis; majoribus cæruleo-viridibus, flavo-viridi marginatis; cauda subtus lutescente; culmine rostri versus basin biangulato.

Near *T. macrorhynchus*, Wagl.; but the under surface, and especially the sides of the breast and belly, have much less yellow; the shoulders and wing-coverts are dull greenish and blue instead of deep black, and only a few of the lesser wing-coverts are of a greenish black; the greater wing-coverts nearest the body are all green, and the yellow margins are much less conspicuous than in the allied species; the outer webs of the primaries and of the greater and middle wing-coverts are green, instead of blue as in *T. macrorhynchus*. The bill also differs, the culmen being much flattened in its basal half, with distinct angular edges, whereas in the allied species it is regularly rounded. Bill deep red; feet dusky olive; iris olive-yellow, with an outer ring nearly white.

Total length 17 inches; wing 9 1/2 inches; bill, to base, 2 1/8 inches, 1 7/8 inch.

Hab. Bouru, Amboyna, and Ceram.

Remarks.--The Amboyna and Ceram specimens have the wing-coverts a little darker than those from Bouru, but they are still sufficiently distinct from *T. macrorhynchus*.

TRICHOGLOSSUS CYANOGRAMMUS.

T. cyanogrammus, Wagl. Mon. Psitt. p. 554.

T. nigrogularis, G. R. Gray, P. Z. S. 1858, p. 183.

Hab. Bouru, Ceram, and all the Papuan Islands. *Remarks.*--On examining specimens from the above-mentioned localities, I can find only slight individual variations among them, not confined to any given locality. The specimens from the Aru Islands (*T. nigrogularis*, G. R. Gray) exactly agree with the rest.

EOS RUBRA, var.

Psittacus borneus, L.

P. ruber, Gm., Wagl. Mon. Psitt. p. 558.

Hab. Bouru, Amboyna, Ceram, and Matabello Islands. The specimens are rather smaller than those from Amboyna, and have more blue on the wing-coverts, and often a greenish tinge on the wings and tail, which makes them agree with the descriptions of *P. borneus* of the old authors. Might not Bouru have been mistaken for Borneo, and thus led to the erroneous name? *Note.*--Besides the preceding five species of *Psittaci*, Bouru possesses also the *Aprosmictus amboinensis*; but as a specimen was not obtained by me, I have not included it in the present list.

HALIASTUR LEUCOSTERNUS.

Haliastur leucosternus, Gould, B. of Austr. i. pl. 4.

Hab. Bouru and the countries eastward.

BAZA REINWARDTII.

Lophotes reinwardtii, Schleg. & Müll. Verh. Ned. t. 5.

Hab. Bouru, the Moluccas, and Timor.

ACCIPITER RUBRICOLLIS. (Pl. IV.)

Supra nigro-plumbeus, subtus albo-cinereus; nucha et colli lateribus late et intense rufis; genis cinereo-plumbeis; gula ventreque albescentibus; remigibus rectricibusque obscure fasciatis.

Above slaty black; beneath very pale ash, shading into nearly pure white on the throat, belly, and under tail-coverts. Back and sides of the neck extending between the shoulders deep red-brown, a lighter shade of which covers the sides of the breast; the wings and tail are crossed by obscure black bands, which on the lighter undersides of the feathers become distinct blackish bands, less visible on the outer tail-feathers. The under wing-coverts and the base and margins of all the quills beneath are of a light rufous-buff. Bill black, at the base plumbeous; cere, orbits, and feet yellow; iris golden yellow.

Length 14 inches; wing 8 1/4 inches; tail 6 1/4 inches; tarsus 2 1/8 inches; middle toe and claw 2 1/8 inches. The young bird is dusky above, with the feathers rufous-margined; beneath creamy white, with broad dusky stripes down each feather.

Hab. Bouru, Ceram, and Gilolo.

Remark.--This bird resembles on its upper surface *A. erythrauchen*, G. R. Gray (P. Z. S. 1860, p. 344), but is very much larger. As the dimensions of that bird are wrongly printed, I will here correct them. Instead of "length 11' 9", wing 8' 9"," as given, it should be, "length 10' 9", wing 6' 9" " [1].

ACCIPITER CRUENTUS.

Astur cruentus, Gould, Birds of Austr. i. pl. 18.

Hab. Bouru and Timor.

ATHENE HANTU.

Rufa, supra rufo-brunnea; gula pallidiore; fronte genisque albescentibus; corpore subtus, cum cauda, rufescente et albescente indistinctissime fasciato; tectricibus alarum inferioribus rufis; remigibus fuscis, pogonio externo rufo; digitis tarsisque setulosis.

Above dark, beneath bright rufous; tail with very indistinct, narrow, paler bars; forehead, cheeks, and chin whitish; under surface indistinctly banded with narrow fasciæ of darker and lighter rufous or whitish; the under tail-coverts barred with rufous and whitish; quills not barred, except close to their bases; under wing-coverts rufous, not barred; third, fourth, and fifth quills equal; tarsi and toes densely clothed with bristles; bill whitish horn-colour; iris yellow; feet (in the living bird) white.

Length 12 inches; wing 8 3/4 inches; tail 5 inches. This species resembles *A. squamipila*, Bp., in its hairy tarsi, but differs in its coloration and proportions; it is one of the "burong hantus" (ghost-birds) of the natives. *Hab.* Bouru.

SCOPS LEUCOSPILUS.

Ephialtes leucospila, G. R. Gray, P. Z. S. 1860, p. 344.

Hab. Bouru and Gilolo.

CAPRIMULGUS MACROURUS.

C. macrourus, Horsf. Linn. Trans. xiii. p. 142.

Hab. Bouru and the whole archipelago.

DENDROCHELIDON MYSTACEUS.

Cypselus mystaceus, Less. Voy. Coquille, Ois. t. 22.

Hab. Bouru, Moluccas, and New Guinea.

Remark.--This is the limit of the range of this fine Tree-Swift to the westward. In the Sula Islands and Celebes it is replaced by *D. wallacii*, Gould.

CACOMANTIS ASSIMILIS.

Cuculus assimilis, G. R. Gray, P. Z. S. 1858, p. 184.

Hab. Bouru.

This specimen seems to agree with that named and described as above; but these small Cuckoos vary so much in their plumage as to render it very difficult to decide. My specimens seem to show that the same species extends over Celebes, the Moluccas, and New Guinea; and it may be probably the same as *C. tymbonomus*, Müll.

EUDYNAMIS RANSOMI.

E. ransomi, Bp. Consp. Gen. Av. p. 101.

Mas ad. *nigro-violaceo nitens; rostro pallide viridi-olivaceo; pedibus plumbeis.*

The female and young male were described by Bonaparte. The adult male is, like others of the genus, entirely shining blue-black; iris crimson.

Total length 18-19 inches; wing 8-8 1/2 inches; bill, to front, 1 1/8 inch.

Hab. Bouru and Ceram.

CENTROPUS MEDIUS.

C. medius, Bp. Consp. Gen. Av. p. 108.

Bill black; feet blackish lead; iris olive-brown. In the immature bird the plumage above is pale rufous, banded and spotted with black; the tail bronzy black, with about sixteen rufous bands; the under surface yellowish, with small dark spots; the thighs and vent dusky; and the bill pale horn.

Length 18-19 inches; wing 7 1/2-8 inches; bill, from gape, 1 1/2 inch.

Hab. Bouru, Ceram, and Gilolo.

TODIRAMPHUS COLLARIS.

Alcedo collaris. Scop.; Sw. Zool. Ill. t. 57.

Hab. Bouru and the whole archipelago.

TODIRAMPHUS SANCTUS.

Halcyon sancta, Vig. & Horsf.; Gould, Birds of Austr. ii. pl. 21.

Hab. Bouru and the islands eastward.

TANYSIPTERA ACIS.

Supra nigra, subtus albo-rufescens, plumis tenuiter nigro marginatis; plumis pilei cæruleo marginatis, superciliis et corona occipitali magis cæruleis; tectricibus alarum minoribus cæruleis; uropygio albo; tectricibus caudæ superioribus albis, rufo tinctis et nigro marginatis; rectricibus mediis elongatis cæruleis, ad basin fusco et albo maculatis, spatulis albis cæruleo marginatis; aliis albis, externe fusco-cæruleo marginatis, interne albo et nigro maculatis; gula albescente; tectricibus caudæ inferioribus albis.

Forehead and crown black, with the feathers blue-margined; a band over the eyes and round the nape brighter blue; ear-coverts, back, and wings deep black, with the lesser wing-coverts blue-margined, margin of the wing blue-tinged; primaries with the outer webs pale-edged towards the tips; under surface of the body pale buff, nearly white on the throat; the feathers of the breast and flanks with blackish lateral edges; rump white, feathers black-edged, the black increasing to the tail-coverts, the last of which have the outer web black; middle tail-feathers blue, with the bases irregularly white-striped, and the spatulate ends white, with bluish margins; lateral tail-feathers white, with blue margins to the outer webs, and irregular dusky markings on the inner webs. Bill orange-red; feet olive; iris dark.

Total length 14 1/2 inches; wing 4 inches. *Hab.* Bouru.

Remarks.-- This interesting addition to the genus *Tanysiptera* is blacker on the upper surface than any of its alhes. It is also remarkable for the buffy tint and black-edged feathers of the under surface,--characters which in the other species are confined to the young birds. My specimen is, however, in fine plumage and condition, and I have little doubt that these characters are distinctive of the adult bird.

Nine species of the genus have now been described; and a careful examination of the fine series of specimens in my collection having convinced me that they can all be clearly characterized, I will add a table of the species.

Table of the Species of Tanysiptera.

I. With a white dorsal spot.	Species.	Habitats.
1. Beneath cinnamon-red	1. sylvia	N. Australia.
2. Beneath white		
A. Tail and upper tail-coverts blue-margined	2. doris	Morty Island.
B. Tail and upper tail-coverts white	3. sabrina	Kaioa Island.

II. No dorsal spot.		
1. Rump red.	4. nympha	New Guinea.
2. Rump white		
A. Ear-coverts and nape black.		
a. Outer tail-feathers black, blue-edged	5. hydrocharis	Aru Island.
b. Outer tail-feathers white, blue-edged	6. acis	Bouru
B. Ear-coverts and nape dark blue.		
a. Eyebrows and nape lighter blue than the crown, terminal tail-coverts black	7. isis	Batchian and Gilolo.
b. Head uniform blue, tail-coverts all white		
a. Back blue-spotted	8. nais	Amboyna, Ceram.
b. Back uniform	9. galatea	New Guinea and Waigiou.

In this table I have altogether left out the Linnæan *Alcedo dea*, because it is possible we may yet obtain certain evidence as to which species it was applied to. The figure in the 'Planches Enluminées' and the careful description of Brisson agree best with *T. sabrina*, G. R. Gray; and I should have little hesitation in placing that name under *T. dea* as a synonym, but that specimens may yet arrive from Ternate--the locality given by the old authors. It is to be remarked, however, that Kaioa Islands, where I obtained *T. sabrina*, is the southernmost of a chain of islets extending up to Ternate, and nowhere more than eight or nine miles apart; so that it is very improbable there should be another species in that island. There can be herefore, I think, but little doubt that *T. sabrina* is but an individual or local variety of the true *Alcedo dea*.

ALCEDO MOLUCCENSIS.

Alcedo moluccensis, Blyth, Journ. As. Soc. Bengal, 1847.

Hab. Bouru, Celebes, and Gilolo.

CEYX CAJELI. (Pl. V.)

Nigra, subtus rufo-lutea; capite et tectricibus alarum punctis parvis cæruleis ornatis; dorso et caudæ tectricibus pallide cæruleis; gula late alba; genis nigris aut tenuiter cæruleo striatis; flexura et margine alarum, colli et frontis maculis lateralibus rufis; rostro pedibusque dilute corallinis.

Above black; beneath rufous yellow; each feather on the head marked with a very small, subtriangular, light-blue spot; on the back and upper tail-coverts the outer half of each feather is whitish blue; chin and throat pure white; a frontal spot over each nostril, a patch behind the ears, and the bend and margin of the wing rufous; ear-coverts black, and the space below them either black or very finely striated with blue; bill and feet pale coral-red; iris dark.

Length 6 inches; wing 2 3/8 inches; bill, from front, 1 3/8 inch.

Hab. Bouru.

Remarks.--This species is very like *C. lepida*; but differs in the very small spots on the head and the stripe on the back being of quite a different blue colour, and also in the scapulars being entirely black, whereas in the other species they are tipped with rich blue. I have named this species after the town or fort of Cajeli in Bouru, to which island this pretty bird is most probably strictly confined.

EURYSTOMUS PACIFICUS.

Coracias pacifica, Lath.

Eurystomus australis, Gould, B. of Austr. ii. pl. 17.

Hab. Bouru and the islands eastward.

PITTA RUBRINUCHA.

Pitta rubrinucha, Wallace, P. Z. S. 1862, p. 187.

Hab. Bouru.

ACROCEPHALUS AUSTRALIS.

Acrocephalus australis, Gould, Birds of Austr. iii. t. 38.

Hab. Bouru.

Remarks.--My specimen agrees exactly with Gould's

figure and description. I did not meet with the species in any other of the islands.

CISTICOLA RUSTICA.

Luteo-rufa; supra plumis medialiter nigris; subtus gula et abdomine medio albescentibus; rectricibus subtus rufo terminatis, macula subapicali nigra.

Rufous yellow; feathers of the head with a black stripe, of the back and wing-coverts black with a rufous margin; quills dusky, the primaries narrowly, the secondaries and tertiaries more broadly rufous-margined; tail pale, rufous-tipped; the two middle feathers rufous, with the central part and towards the apex blackish, the rest black; beneath with the sides of the neck, the breast, the flanks, and the under wing-coverts and tail-coverts pale chestnut, becoming nearly pure white on the throat and the middle of the belly; quills beneath brownish black, narrowly margined with pale rufous towards the base; tail beneath dusky, the feathers with narrow margins and broader tips of pale rufous, and each with a large suhapical black spot; bill dusky above, pale beneath; feet and claws pale yellowish; iris pale olive.

Total length 4 inches; wing 1 7/10 inch; tarsus 5/8 inch. *Hab.* Bouru.

Remark.--Very near *C. lineocapilla*, Gould, with which I had at first placed it; but comparison with a specimen in the British Museum has convinced me of its distinctness from any of the Australian species.

MIMETA BOURUENSIS.

Philedon bouruensis, Quoy & Gaimard, Voy. de l'Astrol. t. 8. f. 2.

Tropidorhynchus buruensis, Bp. Consp. Gen. Av. p. 390.

Cinereo-brunnea, subtus pallidior; facie et auriculis fusco-nigris; capite et gula substriatis; torque nuchali indistincto fulvo-cinereo.

Earthy brown; beneath whitish brown; head a little paler, the feathers marked with a central blackish stripe, and on the nape a narrow paler rufescent band; ear-coverts dusky black; lores and cheeks blackish, mixed with whitish; chin and sides of the throat with a dusky stripe on each feather; primaries outwardly edged with pale rufous; under wing-coverts and margins of all the quills beneath towards the base pale rufous or buff; under tail-coverts with a tinge of buff; rectrices, all but the middle pair, tipped on the inner web with the same colour; bill horny black; feet lead-colour; iris dull red.

Length 9 inches; wing 5 5/8 inches; tail 4 3/4 inches; bill, to front, 1 1/2 inch.

Hab. Bouru (Moluccas).

Remarks.--This curious bird resembles so closely a Honeysucker of the genus *Tropidorhynchus* that it has been figured and described as such, and even escaped the acute eye of Prince Bonaparte, who has given it that place in his 'Conspectus.' But, more singular still, there is a species of true *Tropidorhynchus* inhabiting the same island of Bouru, which so closely resembles this bird that the two can hardly be distinguished, except by a close comparison of the generic characters that separate them. We have here, in fact, a case among birds of that *mimicry* of one species by another belonging to a different group, which Mr. Bates has so well illustrated among the Lepidoptera of S. America (see Linn. Trans. vol. xxiii. p. 495). In this case the Oriole has imitated the Honeysucker; for it has departed from the usual gay colouring of its allies, and is actually the dullest-coloured of its family, while the Honeysucker very much resembles in its coloration other species of the group to which it belongs. The imitation is carried to the minutest particulars: the bare black orbits of the *Tropidorhynchus* are copied by a patch of dusky feathers in the *Mimeta*; the rigid lanceolate feathers on the head of the former are imitated by dark stripes on the broader feathers of the latter; and even the very peculiar ruff of recurved feathers on the nape of the *Tropidorhynchus* has its general effect imitated by a collar of a pale colour in the *Mimeta*. The under and upper surfaces of the two birds are as near as possible of the same tint respectively; and, stranger still, the Oriole has closely copied the mode of flight and the voice of its model; so that in a state of nature the two birds are practically undistinguishable. Most of the species of *Tropidorhynchus* have an elevated keel or protuberance at the base of the bill. In the Bouru bird this is altogether wanting; yet in the *Mimeta* which copies it there is a slight protuberance at the base of the bill, which does not occur in any other species of its genus--almost making us think that some ancestors of the present bird had mimicked a species of *Tropidorhynchus* which possessed the protuberance, and that their descendant, finding himself in the company of a bird without this ornament, was gradually losing it, but had not yet quite done so. It has been observed by Mr. Bates, and is no doubt generally true, that mimicking species are much less abundant than those they copy. In the present instance it seems to be different; for I obtained many specimens of the *Mimeta* before I saw a single *Tropidorhynchus*, though in other islands the latter was generally the most abundant. Perhaps in this case it has carried the imitation to such an extent as actually to gain an advantage over its model in the struggle for existence. This curious instance of mimicry does not stand alone; for in the adjacent island of Ceram, two allied but very distinct species (*Mimeta forsteni* and *Tropidorhynchus subcornutus*) resemble each other with equal accuracy. What peculiar immunity from danger the *Tropidorhynchi* possess, which makes it advantageous for other birds to imitate them, it is not very easy to see. In the case of insects, it seems probable that it is the odour or taste of the imitated species which is unpalatable to insect-eating birds; or, in other cases, like the clear-winged Moths which mimic Hymenoptera, the mimicked species are armed with a sting. In birds it is evident that the bravest, strongest, and best-armed groups should be the subjects of mimicry, and the weakest and most defenceless those which obtain some advantage by imitating them. Now this is certainly the case, for the *Raptores* are the most frequent subjects of imitation--a Parrot (*Strigops*) imitating an Owl, some Curassows of the genus *Ibycter* resembling Hawks (Ibis, vol. ii. p. 223), and Cuckoos frequently resembling Hawks. A species was named by Temminck *Falco cuculoides*; and in all parts of the world the larger grey and banded Cuckoos are mistaken by the natives for Hawks. Cuckoos, however, which are certainly among the weakest and most defenceless of birds, imitate several other groups, especially Gallinaceæ,--for example, *Centropus phasianus* in Australia, and *Carpococcyx radiatus* in Borneo, which latter is terrestrial in its habits, and much resembles the *Euplocami* of the same island. *Eudynamis* also frequently resembles Pigeons, especially the females and young birds, which are banded like *Macropygia*. Among the small Cuculinæ some are very like *Campephagæ* and *Chrysococcyx* has put on the metallic plumage of *Lamprotornis*.

Returning now to *Mimeta* and *Tropidorhynchus*, we have to observe that the former is a smaller, weaker, less active, less noisy, and less pugnacious bird; the feet have a less powerful grasp, and the bill is less acute. The latter has a great variety of loud and piercing notes, which bring its companions to the rescue in time of danger; and I have observed them drive away crows and even hawks which had ventured to perch on a tree where two or three of them were feeding. The *Tropidorhynchus* knows how to take care of himself, and make himself both respected and feared; it would therefore evidently be to the advantage of the more defenceless *Mimeta* to be mistaken for him.

In this instance, as in most others, the imitation is far closer in the living bird than in the dead specimens, and it is a far more satisfactory case of mimicry than any of those which I have alluded to as occurring among birds, and which are more or less general resemblances to another group; while here we have two *species*, each confined to a single island, and each accurately imitated by a bird of a distinct family, with which it has no direct affinities.

I therefore cannot doubt that this is a true case of mimicry, exactly analogous to that so common among insects, and which my friend Mr. Bates has the honour of having first brought under the same general laws which have regulated all variation in the organic world.

CRINIGER MYSTICALIS.

Viridi-olivaceus; subtus flavo-virens; gula crissoque flavescentibus; mento, loris et palpebris flavis; remigum pogonio interno fusco-nigro; cauda immaculata.

Entirely olive-green, more yellow-tinged beneath,

especially on the throat and under tail-coverts; the lores, chin, and eyelids are pure yellow, and also the basal half of the gape-bristles; bill horny black; feet lead-colour; iris red.

Total length 9 inches; wing 4 1/4; bill to gape 1 inch. *Hab.* Bouru.

Remarks.--This species is nearest to *Criniger simplex*, from Gilolo (Ibis, 1862, p. 350); but is at once distinguished by the markings of the face and the remarkable half- yellow gape-bristles.

ARTAMUS LEUCOGASTER.

Lanius leucogaster, Val. Ann. Mus. H. Nat. iv. t. 7. f. 2.

Hab. Bouru and the whole archipelago, from Sumatra to New Guinea.

Remarks.--From the large specimens of N. Celebes to the small ones of Timor and New Guinea there is such a gradation of size in the various islands that it is impossible to separate birds which otherwise agree exactly in form and coloration. *A. papuensis*, Bp. Consp. p. 344, will have to be considered as a very slight local variety of the present bird.

HIRUNDO JAVANICA.

Hirundo javanica, Lath., Temm. Pl. Col. 83. f. 2.

Hab. Bouru and the islands westward.

MYIAGRA GALEATA.

Myiagra galeata, G. R. Gray, P. Z. S. 1860, p. 352.

Hab. Bouru and the Moluccas.

Remarks.--The only two specimens procured are ashy above, with faint signs of glossy blue and rufous white beneath; they probably show the immature plumage of the species, of which I have specimens from Ceram and the small islands east of it, and also from Morty, north of Gilolo.

MONARCHA LORICATA. (Pl. VI.)

Nigro-chalybea, subtus alba; mento gulaque squamatis, nigro-chalybeis; cauda alba, rectricibus mediis nigris, duabus utrinque juxta medium nigro terminatis; rostro pedibusque cæruleo-plumbeis.

Blue-black; beneath pure white, except the throat, which is covered with scaly feathers of a rich metallic blue-black; this colour meets the black of the upper parts at the angle of the mouth, and extends in an oval shield to the bottom of the neck; under wing-coverts white; tail with the three lateral feathers on each side entirely white, the next two black-tipped, and the middle pair entirely black, with occasionally some white touches on the outer webs; bill and feet lead-blue; iris dark.

The sexes are alike; in the young bird the white is replaced by pale reddish brown, and the black by fuscous brown.

Total length 7 inches; wing 3 1/2 inches.

Hab. Bouru Islands (Moluccas).

Remarks.--This beautiful species is nearly allied to *M.*

leucura of Mr. G. R. Gray, which I sent from the Ké Islands, east of Ceram.

RHIPIDURA TRICOLOR.

Musicapa tricolor, Vieill. N. Dict. Hist. Nat. xvi. p. 490.

Hab. Bouru, Moluccas, and New Guinea.

RHIPIDURA BOURUENSIS.

Fusco-plumbea; capite nigro, ventre pallide rufo, alis caudaque fuscis; gula albescente, pectoris maculis elongatis albis; stria supraoculari occulta, alba; tectricibus majoribus pallide terminatis, remigibus ultimis pallide marginatis; rectricum duarum externarum pogonio externo rufo-albo.

Dusky lead-colour, deepening on the head to black; wings and tail dusky brown; feathers of the throat somewhat decomposed, with the outer half white; those of the breast with an elongate oval white spot on each feather; middle of the belly, the vent, and under tail-coverts pale rufous; over the eye is a silvery-white mark, only visible when the feathers are raised; the under wing-coverts are tipped with pale rufous, the outer row with white; the greater wing-coverts above have the extreme apex whitish, the tertiary quills have a very narrow pale-rufous edging; the tail is immaculate, with the exception of the two outer quills, which have their outer web for its whole length rusty white; bill black; feet dusky; iris dark.

Length 7 inches; wing 3 3/8 inches; tail 3 1/2 inches; bill, to front, 1/2 inch.

Hab. Bouru.

Remarks.--I have named this species after the island it inhabits, because the allied forms from the surrounding islands being already known, there is every probability of its never being found anywhere else.

PACHYCEPHALA LINEOLATA.

Pachycephala lineolata, Wallace, P. Z. S. 1862, p. 341.

Hab. Bouru and Sula Islands.

PACHYCEPHALA RUFESCENS.

Pachycephala rufescens, Wallace, P. Z. S. 1862, p. 341.

Hab. Bouru and Sula Islands.

PACHYCEPHALA CLIO.

Pachycephala clio, Wallace, P. Z. S. 1862, p. 341.

Hab. Bouru and the Sula Islands.

Remarks.--The Bouru specimens have a more yellow tinge on the back, and the black pectoral band is generally broader than in those from Sula. I may here observe that the fine species from Batchian and Ternate, included in Mr. G. R. Gray's list of Molucca birds as *P. melanura*, Gould, is quite distinct from that species, and may be recognized by its black chin and upper tail-coverts, and narrow black crescent on the breast entirely disconnected from the black head, and also by its much larger size. We have therefore in the Moluccas four species of *Pachycephala* allied to the *pectoralis* and *melanura* of Australia, viz. *P. macrorhyncha*, Strickl., in Amboyna and Ceram, *P. calliope*, Bp., in Timor, *P. clio* in Bouru and Sula, and *P. mentalis*, n. s., in Batchian, Ternate, and Gilolo[2].

DICRURUS AMBOINENSIS.

Dicrurus amboinensis, G. R. Gray, P. Z. S. 1860, p. 354.

Hab. Bouru.

Remark.--The specimens are rather larger and better-coloured than those from Amboina and Ceram, but otherwise agree with them.

CAMPEPHAGA MARGINATA.

Supra plumbea, subtus albo-cinerea; loris fusco-nigris; tectricibus caudæ et alarum inferioribus albis; remigibus et tectricibus alarum majoribus nigris, albo marginatis; rectricibus extimis utrinque tribus albo terminatis.

Bluish lead-colour above, ashy white beneath; base of wings beneath and under tail-coverts white; wings and tail black; primaries white-margined on the inner web near the base; secondaries, tertiaries, and greater wing-coverts white-margined towards the points; middle tail-feathers ashy, with a black tip, outer ones with the outer margin and tip ashy, the next two with diminishing ashy tips; bill and feet black; iris dark.

Total length 8 1/2 inches; wing 4 1/4 inches.

Hab. Bouru.

Remarks.--This species somewhat resembles *C. plumbea*, but is smaller and paler beneath, and the bill is more slender.

TROPIDORHYNCHUS BOURUENSIS.

Cinereo-brunneus, subtus pallide cinereus; gula et capitis lateribus plumis subrigidis subsericeis; alis caudaque subtus fuscis; rectricum lateralium utrinque duarum apicibus tenuiter fulvescentibus; facie nuda nigra; protuberantia ad basin rostri nulla.

Above ashy brown; head somewhat paler, with lanceolate feathers, the stems of which are black; beneath pale ashy; the plumes of the throat and upper part of the breast and the marginal feathers of the head and face somewhat rigid, of a silky lustre, and with darker stems; quills dusky, with the inner margins of a paler fulvous tinge; tail uniform dusky, the two outer feathers on each side with the apex on the inner side of paler fulvous colour; orbits and cheeks bare, black; bill black, without any protuberance at the base; feet pale lead-colour; iris light olive.

Length 14 1/2 inches; wing 6 inches; tail 5 1/2 inches; bill, to front, 1 3/4 inch.

Hab. Bouru.

Remarks.--This species is the subject of imitation by a bird of quite distinct family (Oriolidæ), as fully explained under *Mimeta bouruensis*, which bird is the *Tropidorhynchus bouruensis*, Bp., ex Lesson.

ZOSTEROPS CHLORIS.

Zosterops chloris, Bp. Consp. Gen. Av. p. 398.

Hab. Bouru, Ternate, and Banda. "Iris pale brown; bill dusky black above; beneath and feet lead-colour."

DICÆUM ERYTHROTHORAX

Diceum erythrothorax, Less. Voy. Coquille, Ois. t. 30. f. 1, 2.

Hab. Bouru.

Remark.--An allied species to this occurs in Ceram, of which I give the description in a note[3].

NECTARINIA PROSERPINA.

Purpureo-nigra velutina; capite viridi-chalybeo; gula purpureo-violacea metallica; crisso, tectricibus caudæ superioribus et alarum minoribus purpureo-cyaneis; remigibus fusco-nigris; cauda elongata, rectricibus duabus mediis purpureo marginatis. ♀ . *Supra olivaceo-viridis, subtus flavescens; capite pectoreque cinereis; cauda fuscescenti-nigra, apice pallida.*

Rich velvety purple-black; crown greenish steel-blue;

throat richly scaled with violet-purple; wings, with the lesser coverts only, the rump, and upper tail-coverts metallic-blue; two middle tail-feathers margined on both sides with purple; wings and tail fuscous black.

Female.--Above olive-green; the crown and nape dark ash, each feather having a central dusky spot; beneath pale olive-yellow, the throat and breast light ash; quills dusky, with an outer margin of olive-yellow; tail purplish-black, the feathers margined on the outer web with olive-green, and a whitish spot on the inner web at the apex, increasing in size from the middle to the outer feathers.

Length 5 inches; wing 2 1/3 inches; tail 1 1/2 inch; bill, from front, 3/4 inch.

Hab. Bouru.

Remark.--This beautiful species is like *N. aspasia*, but differs in its middle and greater wing-coverts being purple-black and not metallic, and in the longer tail.

NECTARINIA ZENOBIA.

Cinnyris zenobia, Less. Voy. Coq.

C. clementiæ, Less. Man. d'Orn. ii. p. 40.

Hab. Bouru, Amboyna, Ceram, and Ké Islands.

CALORNIS OBSCURA.

Lamprotornis obscura, Bp. (ex Forsten) Consp. Gen. Av. 417.

Hab. Bouru and the other Moluccas.

MUNIA MOLUCCA.

Loxia molucca, L.; Pl. Enl. 139. 2.

Hab. Bouru and the Moluccas.

TRERON AROMATICA.

Columba aromatica, Gm.

C. viridis amboinensis, Br. Orn. i. p. 146; Pl. Enl. 163 (fig. pessima).

Bill, cere, and eyelids pale dull blue, tip of the bill becoming yellow in dry specimens; iris white; feet dusky purple.

Total length 11 1/2 inches; wing 6 inches.

Hab. Bouru, and probably Amboyna and Ceram.

Remarks.--Brisson's description of this species is most accurate, and, with the bird before one, cannot be mistaken. The figure in the 'Planches Enluminées' is abominable, but no doubt applies only to this bird. Gmelin copies Brisson; but makes an error which would prevent one recognizing the bird, in saying that "the *upper* tail-coverts are sordid white," instead of the *lower*. This bird is the true *Treron aromatica* (as being an inhabitant of the Spice Islands), a name which has been applied to birds of distinct species from India, Sumatra, and the Philippine Islands. It is easily distinguished from all its allies by having the top of the head ashy blue, not reaching below or even to the eyes, by the broad yellow bands on the wings, and by the under tail-coverts being nearly pure white in *both sexes*.

CARPOPHAGA MELANURA.

Carpophaga melanura ?, G. R. Gray, P. Z. S. 1860, p. 361.

Hab. Bouru and all the Moluccas.

Remarks.--This species is certainly distinct and peculiar to the Moluccas, *C. luctuosa* being found only in Celebes on the west, and *C. bicolor* in the Papuan Islands to the east of it. Bill greenish horn-colour; tip greenish yellow; feet lead-colour; iris nearly black.

CARPOPHAGA PERSPICILLATA, var.

Columba perspicillata, Temm. Pl. Col. 246.

Hab. Bouru, Batchian, Gilolo, and Waigiou Islands. *Remarks.*--The true *C. perspicillata* of Temminck is probably that of the islands of Ceram and Amboyna, which has the head and neck of nearly the same whitish ash as the under surface of the body, and the quills of a powdery-ash tint; whereas in the specimens from the Northern Moluccas and Bouru the head and sides of the neck are slate-colour, the throat and breast slaty ash, and the quills purple-black, with a slight ashy tinge. The bill is bluish, pale at the tip, and red at the base; the feet pale purple, and the iris brown-black. This variety is constant and easily distinguishable, and will probably be considered a distinct species by many naturalists; and it is only the absence of any perceptible difference in the form or proportions, or of any definite markings which can be more clearly characterized than shades of colour, which prevents me classing it as such.

PTILONOPUS PRASINORRHOUS.

Ptilonopus prasinorrhous, G. R. Gray, P. Z. S. 1858, p. 185.

Bill and its base, as far as the eye, gamboge-yellow; iris orange-brown, with an inner ring of yellow; feet dull purple.

Length 9 1/2 inches; wing 4 3/4 inches.

Female entirely green; crown of head very rich green; underside rather duller; under tail-coverts yellow-margined. *Hab.* Bouru, Matabello, Goram, and Ké Islands, belonging to the Molucca group; also Mysol and Waigiou, of the New Guinea group.

PTILONOPUS VIRIDIS.

Columba viridis, L., Pl. Enl. 142.

Bill yellow, the base red; iris yellowish red; orbits yellow; feet red.

Total length 9 inches. The male and female are alike. *Hab.* Bouru, Amboyna, Ceram, and Goram.

MACROPYGIA AMBOINENSIS.

Columba amboinensis, L., Bp. Consp. Gen. Av. ii. p. 56.

Hab. Bouru and the other Moluccas.

"Bill black; iris pearly white, with an outer ring of pink; feet coral-red."

CHALCOPHAPS MOLUCCENSIS.

Chalcophaps moluccensis, G. R. Gray, P. Z. S. 1862, p. 345.

Hab. Bouru, Sula Islands, and the Moluccas.

MEGAPODIUS WALLACII.

Megapodius wallacei, G. R. Gray, P. Z. S. 1860, p. 362, pl. 171.

This species differs somewhat in its habits from the other members of the family found in the Malay Islands. It resides generally in the hilly districts of the interior, like *Megacephalon maleo*, and, like that species, comes down to the beach to deposit its eggs; but instead of scratching a hole for them and covering it up again, the bird burrows into the sand to the depth of 3 or 4 feet obliquely downwards, and deposits its egg at the bottom. It then loosely covers up the mouth of its hole; and is said by the natives to obliterate and disguise, by innumerable tracks and scratches, its own footmarks leading to the hole. Its offspring is then left to make its way into the world as it best can. The only specimen

I obtained here was caught on the beach, at the mouth of its burrow, early one morning. Its wing was broken and wounded at the outer joint, as if it had been attacked by some small animal when in its burrow, probably a rat.

Hab. Bouru, Gilolo, and Ternate.

MEGAPODIUS FORSTENI.

Megapodius forsteni, Gray & Mitch. Gen. of Birds, iii. pl. 124.

Hab. Bouru, Amboyna, and Ceram.

This bird deposits its eggs in a heap of rubbish collected in low places near the sea. It is seminocturnal in its habits, making a loud wailing cry, which is often heard at night and about daybreak.

GLAREOLA GRALLARIA.

Glareola grallaria, Temm.

G. australis, Leach, Linn. Trans. xiii. pl. 14.

Hab. Bouru and Australia.

ESACUS MAGNIROSTRIS.

Charadrius magnirostris, Lath., Temm. Pl. Col. 387.

Hab. Bouru, Celebes, and New Guinea.

NUMENIUS UROPYGIALIS.

Numenius uropygialis, Gould, B. of Austr. vi. pl. 43.

Hab. Bouru, the Moluccas, and New Guinea.

BUTORIDES JAVANICUS.

Ardea javanica, Horsf. Linn. Trans, xiii. p. 190.

Hab. Bouru and the whole archipelago.

ARDETTA FLAVICOLLIS.

Ardea flavicollis, Lath. Ind. Orn. ii. p. 701.

Hab. Bouru, and from India to Australia.

NYCTICORAX CALEDONICUS.

Ardea caledonica, Gm.: Gould, B. of Austr. 6. t. 63.

Hab. Bouru, Moluccas, and Australia.

GALLINULA FRONTATA.

Fusco-plumbeo-nigra; dorso alisque olivascentibus; cauda nigra; tectricibus caudæ inferioribus lateralibus albis, mediis nigris; rostro rubro, apice abrupte luteo; clypeo frontali magno, dilatato, supercilia attingente; pedibus rubris, fusco articulatis, tibiis subtus olivaceis.

Very near *G. tenebrosa*, Gould, but distinguishable from that species by the differently-coloured back and wings, which are olivaceous brown instead of black, the rather slenderer bill, the very large frontal plate, and the more uniform-coloured legs, the joints of the tibiæ and tarsi being dusky olive, the median line of the tibiæ beneath olive-green, the tarsi beneath dusky lead-colour; the lateral under tail-coverts pure white, the middle ones black; the rest of the plumage of a dusky lead-colour, deepening on the top of the head and neck to nearly black, and on the breast tinged with brown; wings brownish olive; tail black.

Total length 13 inches; wing 7 inches; bill, from back of frontal plate, 2 inches.

Hab. Bouru.

RALLINA PHILIPPENSIS.

Rallus philippensis, L.

Hab. Bouru and the islands eastward.

DENDROCYGNA GUTTULATA.

Anas guttulata, Temm.

Hab. Bouru, Ceram, and Celebes.

TADORNA RADJAH.

Anas radjah. Less. & Garn. Voy. Coq. Zool. pl. 49.

Eyelids yellow; iris milk-white; bill and feet white; claws dusky white; nostrils blackish.

Total length 21 inches.

Hab. Bouru, the Moluccas, and New Guinea.

Remarks.--Lesson describes the bill and feet of this bird as *red.* In dried specimens they become of a dull reddish white; but in the living bird are entirely white. Lesson's specimen was obtained in Bouru.

PODICEPS TRICOLOR.

Podiceps tricolor, G. R. Gray, P. Z. S. 1860, p. 366.

Bill black; base of the lower mandible lemon-yellow, which extends up towards the eye.

Hab. Bouru and the Moluccas.

FALCO RUBRICOLLIS

CEYX CAJELI

MONARCHA LORICATA

List of Birds

Notes Appearing in the Original Work

1. Since reading this paper, I have seen Professor Schlegel's 'Catalogue of the Birds in the Leyden Museum,' part 1, in which (Astures, p. 39) he describes this bird under the name of *Nisus cirrhocephalus ceramensis*, which seems to be equivalent to making it a variety of *N. cirrhocephalus*. Considering, however, the bird to be a very good species, I should at once have adopted Professor Schlegel's name *ceramensis*, had I not obtained the bird in other localities than Ceram. The *Raptores* having so generally an extensive range renders the application of territorial specific names less advisable in their case than in that of the *Passeres*. My own rule is only to apply the name of a country as specific name when the surrounding districts are known to possess their peculiar representative species, in which case it amounts almost to a certainty that the new bird is similarly restricted in range.

2. PACHYCEPHALA MENTALIS, n. s. *P. flavo-olivacea; capite, genis mentoque nigris; gula late alba; lunula pectorali nigra; subtus cum torque nuchali vivide flava; cauda ejusque tectricibus superioribus nigris; remigibus fusco-nigris, primariis olivaceo limbatis, aliis tectricibusque alarum flavo-olivaceo marginatis; rostro nigro, pedibus fusco-olivaceis.* Long. 7, alar. 3.7, caudæ 2.10, poll. et duodecim.

Hab. Ins. Batchian et Gilolo.

This may be *Tardus armillaris,* Temm., or *Lanius cucullatus,* Licht. (Bp. Consp. p. 328); but I can find no descriptions of those species, and therefore give this bird a name descriptive of a peculiarity confined to it. *Laniarius albicollis,* Vieill., is different from this and apparently from any other described species.

3. DICÆUM VULNERATUM.

Supra æneo-fuscum subtus cinereum; abdomine albescente; macula parva pectorali tectricibusque caudæ superioribus rubris. Fem. *immaculata.*

Above dark fuscous brown, with a bronzy tinge; beneath light ashy, becoming nearly white on the belly and vent; a small round patch on the breast and the upper tail-coverts bright red; under wing-coverts and sides of breast white; bill black, bluish at the base; feet black. The female (?) or young bird is rather lighter on the upper and darker on the under surface than the male, has no red spot on the breast, and the upper tail-coverts are reddish olivaceous. Total length 3 inches; wing 2 inches.

Hab. Ceram.

MONKEYS

BY

ALFRED RUSSEL WALLACE

British Library Cataloguing-in-Publication Data
A catalogue record for this book is available from the
British Library

Alfred Russel Wallace

Alfred Russel Wallace was born on 8th January 1823 in the village of Llanbadoc, in Monmouthshire, Wales.

At the age of five, Wallace's family moved to Hertford where he later enrolled at Hertford Grammar School. He was educated there until financial difficulties forced his family to withdraw him in 1836. He then boarded with his older brother John before becoming an apprentice to his eldest brother, William, a surveyor. He worked for William for six years until the business declined due to difficult economic conditions.

After a brief period of unemployment, he was hired as a master at the Collegiate School in Leicester to teach drawing, map-making, and surveying. During this time he met the entomologist Henry Bates who inspired Wallace to begin collecting insects. He and bates continued exchanging letters after Wallace left teaching to pursue his surveying career. They corresponded on prominent works of the time such as Charles Darwin's *The Voyage of the Beagle* (1839) and Robert Chamber's *Vestiges of the Natural History of Creation* (1844).

Wallace was inspired by the travelling naturalists of the day and decided to begin his exploration career collecting

specimens in the Amazon rainforest. He explored the Rio Negra for four years, making notes on the peoples and languages he encountered as well as the geography, flora, and fauna. On his return voyage his ship, Helen, caught fire and he and the crew were stranded for ten days before being picked up by the Jordeson, a brig travelling from Cuba to London. All of his specimens aboard Helen had been lost.

After a brief stay in England he embarked on a journey to the Malay Archipelago (now Singapore, Malaysia, and Indonesia). During this eight year period he collected more than 126,000 specimens, several thousand of which represented new species to science. While travelling, Wallace refined his thoughts about evolution and in 1858 he outlined his theory of natural selection in an article he sent to Charles Darwin. This was published in the same year along with Darwin's own theory. Wallace eventually published an account of his travels *The Malay Archipelago* in 1869, and it became one of the most popular books of scientific exploration in the 19th century.

Upon his return to England, in 1862, Wallace became a staunch defender of Darwin's landmark work *On the Origin of Species* (1859). He wrote responses to those critical of the theory of natural selection, including 'Remarks on the Rev. S. Haughton's Paper on the Bee's Cell, And on the Origin of Species' (1863) and 'Creation by Law' (1867). The former of these was particularly pleasing to Darwin. Wallace also

published important papers such as 'The Origin of Human Races and the Antiquity of Man Deduced from the Theory of 'Natural Selection" (1864) and books, including the much cited *Darwinism* (1889).

Wallace made a huge contribution to the natural sciences and he will continue to be remembered as one of the key figures in the development of evolutionary theory.

Wallace died on 7th November 1913 at the age of 90. He is buried in a small cemetery at Broadstone, Dorset, England.

MONKEYS

POPULAR SCIENCE MONTHLY VOLUME 21 MAY 1882

IF the skeletons of an orang-outang and a chimpanzee be compared with that of a man, there will be found to be the most wonderful resemblance, together with a very marked diversity. Bone for bone, throughout the whole structure, will be found to agree in general form, position, and function, the only absolute differences being that the orang has nine wrist-bones, whereas man and the chimpanzee have but eight; and the chimpanzee has thirteen pairs of ribs, whereas the orang, like man, has but twelve. With these two exceptions, the differences are those of shape, proportion, and direction only, though the resulting differences in the external form and motions are very considerable. The greatest of these are, that the feet of the anthropoid or man-like apes, as well as those of all monkeys, are formed like hands, with large opposable thumbs fitted to grasp the branches of trees, but unsuitable for erect walking, while the hands have weak small thumbs but very long and powerful fingers, forming a hook rather than a hand, adapted for climbing up trees and suspending the whole weight from horizontal branches. The

almost complete identity of the skeleton, however, and the close similarity of the muscles and of all the internal organs, have produced that striking and ludicrous resemblance to man which every one recognizes in these higher apes and, in a less degree, in the whole monkey tribe; the face and features, the motions, attitudes, and gestures being often a strange caricature of humanity. Let us, then, examine a little more closely in what the resemblance consists, and how far, and to what extent, these animals really differ from us.

Besides the face, which is often wonderfully human although the absence of any protuberant nose gives it often a curiously infantile aspect monkeys, and especially apes, resemble us most closely in the hand and arm. The hand has well-formed fingers with nails, and the skin of the palm is lined and furrowed like our own. The thumb is, however, smaller and weaker than ours, and is not so much used in taking hold of anything. The monkey's hand is, therefore, not so well adapted as that of man for a variety of purposes, and can not be applied with such precision in holding small objects, while it is unsuitable for performing delicate operations such as tying a knot or writing with a pen. A monkey does not take hold of a nut with its forefinger and thumb as we do, but grasps it between the fingers and the palm in a clumsy way, just as a baby does before it has acquired the proper use of its hand. Two groups of monkeys one in Africa and one in South America have no thumbs on

their hands, and yet they do not seem to be in any respect inferior to other kinds which possess it. In most of the American monkeys the thumb bends in the same direction as the fingers, and in none is it so perfectly opposed to the fingers as our thumbs are; and all these circumstances show that the hand of the monkey is, both structurally and functionally, a very different, and very inferior organ to that of man, since it is not applied to similar purposes, nor is it capable of being so applied.

When we look at the feet of monkeys we find a still greater difference, for these have much larger and more opposable thumbs, and are, therefore, more like our hands; and this is the case with all monkeys, so that even those which have no thumbs on their hands, or have them small and weak and parallel to the fingers, have always large and well-formed thumbs on their feet. It was on account of this peculiarity that the great French naturalist Cuvier named the whole group of monkeys Quadrumana, or four-handed animals, because, besides the two hands on their fore-limbs, they have also two hands in place of feet on their hind-limbs. Modern naturalists have given up the use of this term, because they say that the hind extremities of all monkeys are really feet, only these feet are shaped like hands; but this is a point of anatomy, or rather of nomenclature, which we need not here discuss.

Let us, however, before going further, inquire into the purpose and use of this peculiarity, and we shall then see that it is simply an adaptation to the mode of life of the animals which possess it. Monkeys, as a rule, live in trees, and are especially abundant in the great tropical forests. They feed chiefly upon fruits, and occasionally eat insects and birds' eggs, as well as young birds, all of which they find in the trees; and, as they have no occasion to come down to the ground, they travel from tree to tree by jumping or swinging, and thus pass the greater part of their lives entirely among the leafy branches of lofty trees. For such a mode of existence, they require to be able to move with perfect ease upon large or small branches, and to climb up rapidly from one bough to another. As they use their hands for gathering fruit and catching insects or birds, they require some means of holding on with their feet, otherwise they would be liable to continual falls, and they are able to do this by means of their long finger-like toes and large opposable thumbs, which grasp a branch almost as securely as a bird grasps its perch. The true hands, on the contrary, are used chiefly to climb with, and to swing the whole weight of the body from one branch or one tree to another, and for this purpose the fingers are very long and strong, and in many species they are further strengthened by being partially joined together, as if the skin of our fingers grew together as far as the knuckles. This shows that the separate action of the fingers, which is so

important to us, is little required by monkeys, whose hand is really an organ for climbing and seizing food, while their foot is required to support them firmly in any position on the branches of trees, and for this purpose it has become modified into a large and powerful grasping hand.

Another striking difference between monkeys and men is, that the former never walk with ease in an erect posture, but always use their arms in climbing or in walking on all-fours like most quadrupeds. The monkeys that we see in the streets, dressed up and walking erect, only do so after much drilling and teaching, just as dogs may be taught to walk in the same way; and the posture is almost as unnatural to the one animal as it is to the other. The largest and most man-like of the apes the gorilla, chimpanzee, and orang-outang also walk usually on all-fours; but in these the arms are so long and the legs so short that the body appears half erect when walking; and they have the habit of resting on the knuckles of the hands, not on the palms like the smaller monkeys, whose arms and legs are more nearly of an equal length, which tends still further to give them a semi-erect position. Still, they are never known to walk of their own accord on their hind-legs only, though they can do so for short distances, and the story of their using a stick and walking erect by its help in the wild state is not true. Monkeys, then, are both four-handed and four-footed beasts; they possess four hands formed very much like our hands, and capable of

picking up or holding any small object in the same manner; but they are also four-footed, because they use all four limbs for the purpose of walking, running, or climbing; and, being adapted to this double purpose, the hands want the delicacy of touch and the freedom as well as the precision of movement which ours possess. Man alone is so constructed that he walks erect with perfect ease, and has his hands free for any use to which he wishes to apply them; and this is the great and essential bodily distinction between monkeys and men.

We will now give some account of the different kinds of monkeys and the countries they inhabit.

The Different Kinds of Monkeys and the Countries they inhabit. Monkeys are usually divided into three kinds apes, monkeys, and baboons; but these do not include the American monkeys, which are really more different from all those of the Old World than any of the latter are from each other. Naturalists, therefore, divide the whole monkey-tribe into two great families, inhabiting the Old and the New Worlds respectively; and, if we learn to remember the kind of differences by which these several groups are distinguished, we shall be able to understand something of the classification of animals, and the difference between important and unimportant characters.

Taking first the Old World groups, they may be thus defined: apes have no tails; monkeys have tails, which are

usually long; while baboons have short tails, and their faces, instead of being round and with a man-like expression as in apes and monkeys, are long and more dog-like. These differences are, however, by no means constant, and it is often difficult to tell whether an animal should be classed as an ape, a monkey, or a baboon. The Gibraltar ape, for example, though it has no tail, is really a monkey, because it has callosities, or hard pads of bare skin on which it sits, and cheek-pouches in which it can stow away food; the latter character being always absent in the true apes, while both. are present in most monkeys and baboons. All these animals, however, from the largest ape to the smallest monkey, have the same number of teeth as we have, and they are arranged in a similar manner, although the tusks, or canine teeth, of the males are often large, like those of a dog. The American monkeys, on the other hand, with the exception of the marmosets, have four additional grinding-teeth (one in each jaw on either side), and none of them have callosities or cheek-pouches. They never have prominent snouts like the baboons; their nostrils are placed wide apart and open sideways on the face; the tail, though sometimes short, is never quite absent; and the thumb bends the same way as the fingers, is generally very short and weak, and is often quite wanting. We thus see that these American monkeys differ in a great number of characters from those of the Eastern hemisphere; and they have this further peculiarity,

that many of them have prehensile or grasping tails, which are never found in the monkeys of any other country. This curious organ serves the purpose of a fifth hand. It has so much muscular power that the animal can hang by it easily with the tip curled round a branch, while it can also be used to pick up small objects with almost as much ease and exactness as an elephant's trunk. In those species which have it most perfectly formed it is very long and powerful, and the end has the underside covered with bare skin, exactly resembling that of the finger or palm of the hand, and apparently equally sensitive. One of the common kinds of monkeys that accompany street organ-players has a prehensile tail, but not of the most perfect kind; since in this species the tail is entirely clad with hair to the tip, and seems to be used chiefly to steady the animal when sitting on a branch by being twisted round another branch near it. The statement is often erroneously made that all American monkeys have prehensile tails; but the fact is that rather less than half the known kinds have them so, the remainder having this organ either short and bushy or long and slender, but entirely without any power of grasping. All prehensile-tailed monkeys are American, but all American monkeys are not prehensile-tailed.

By remembering these characters it is easy, with a little observation, to tell whether any strange monkey comes from America or from the Old World. If it has bare seat-pads,

or if when eating it fills its mouth till its cheeks swell out like little bags, we may be sure it comes from some part of Africa or Asia; while, if it can curl up the end of its tail so as to take hold of anything, it is certainly American. As all the tailed monkeys of the Old World have seat-pads (or ischial callosities as they are called in scientific language), and as all the American monkeys have tails, but no seat-pads, this is the most constant external character by which to distinguish them; and, having done so, we can look for the other peculiarities of the American monkeys, especially the distance apart of the nostrils and their lateral position.

The whole monkey-tribe is especially tropical, only a few kinds being found in the warmer parts of the temperate zone. One inhabits the Rock of Gibraltar, and there is one very like it in Japan, and these are the two monkeys which live farthest from the equator. In the tropics they become very abundant, and increase in numbers and variety as we approach the equator, where the climate is hot, moist, and equable, and where flowers, fruits, and insects are to be found throughout the year. Africa has about fifty-five different kinds, Asia and its islands about sixty, while America has one hundred and fourteen, or almost exactly the same as Asia and Africa together. Australia and its islands have no monkeys, nor has the great and luxuriant Island of New Guinea, whose magnificent forests seem so well adapted for them. We will now give a short account of the different

kinds of monkeys inhabiting each of the tropical continents.

Africa possesses two of the great man-like apes the gorilla and the chimpanzee, the former being the largest ape known, and the one which, on the whole, perhaps most resembles man, though its countenance is less human than that of the chimpanzee. Both are found in West Africa, near the equator, but they also inhabit the interior wherever there are great forests; and Dr. Schweinfurth states that the chimpanzee inhabits the country about the sources of the Shari River, in 28 east longitude and 4 north latitude.

The long-tailed monkeys of Africa are very numerous and varied. One group has no cheek-pouches and no thumb on the hand, and many of these have long, soft fur of varied colors. The most numerous group are the guenons, rather small, long-tailed monkeys, very active and lively, and often having their faces curiously marked with white or black, or ornamented with whiskers or other tufts of hair; and they all have large cheek-pouches and good-sized thumbs. Many of them are called green monkeys, from the greenish-yellow tint of their fur, and most of them are well-formed, pleasing animals. They are found only in tropical Africa.

The baboons are larger, but less numerous. Thev resemble dogs in the general form and the length of the face or snout, but they have hands with well-developed thumbs on both the fore and hind limbs; and this, with something in the expression of the face, and their habit of sitting up and using

their hands in a very human fashion, at once shows that they belong to the monkey-tribe. Many of them are very ugly, and in their wild state they are the fiercest and most dangerous of monkeys. Some have the tail very long, others of medium length, while it is sometimes reduced to a mere stump, and all have large cheek-pouches and bare seat-pads. They are found all over Africa, from Egypt to the Cape of Good Hope; while one species, called the hamadryas, extends from Abyssinia across the Red Sea into Arabia, and is the only baboon found out of Africa. This species was known to the ancients, and it is often represented in Egyptian sculptures, while mummies of it have been found in the catacombs. The largest and most remarkable of all the baboons is the mandrill of West Africa, whose swollen and hog-like face is ornamented with stripes of vivid blue and scarlet. This animal has a tail scarcely two inches long, while in size and strength it is not much inferior to the gorilla. These large baboons go in bands, and are said to be a match for any-other animals in the African forests, and even to attack and drive away the elephants from the districts they inhabit.

Turning now to Asia, we have first one of the best known of the large man-like apes the orang-outang, found only in the two large islands, Borneo and Sumatra. The name is Malay, signifying "man of the woods," and it should be pronounced orang-ootang, the accent being on the first syllable of both words. It is a very curious circumstance that,

whereas the gorilla and chimpanzee are both black, like the negroes of the same country, the orang-outang is red or reddish-brown, closely resembling the color of the Malays and Dyaks who live in the Bornean forests. Though very large and powerful, it is a harmless creature, feeding on fruit, and never attacking any other animal except in self-defense. A full-grown male orang-outang is rather more than four feet high, but with a body as large as that of a stout man, and with enormously long and powerful arms.

Another group of true apes inhabit Asia and the larger Asiatic islands, and are in some respects the most remarkable of the whole family. These are the gibbons, or long-armed apes, w^hich are generally of small size and of a gentle disposition, but possessing the most wonderful agility. In these creatures the arms are as long as the body and legs together, and are so powerful that a gibbon will hang for hours suspended from a branch, or swing to-and-fro, and then throw itself a great distance through the air. The arms, in fact, completely take the place of the legs for traveling. Instead of jumping from bough to bough, and running on the branches, like other apes and monkeys, the gibbons move along while hanging suspended in the air, stretching their arms from bough to bough, and thus going hand over hand as a very active sailor will climb along a rope. The strength of their arms is, however, so prodigious, and their hold so sure, that they often loose one hand before they have caught

a bough with the other, thus seeming almost to fly through the air by a series of swinging leaps; and they travel among the network of interlacing boughs a hundred feet above the earth with as much ease and certainty as we walk or run upon level ground, and with even greater speed. These little animals scarcely ever come down to the ground of their own accord; but, when obliged to do so, they run along almost erect, with their long arms swinging round and round, as if trying to find some tree or other object to climb upon. They are the only apes who naturally walk without using their hands as well as their feet; but this does not make them more like men, for it is evident that the attitude is not an easy one, and is only adopted because the arms are habitually used to swing by, and are therefore naturally held upward instead of downward, as they must be when walking on them.

The tailed monkeys of Asia consist of two groups, the first of which have no cheek-pouches, but always have very long tails. They are true forest monkeys, very active, and of a shy disposition. The most remarkable of these is the long-nosed monkey of Borneo, which is very large, of a pale-brown color, and distinguished by possessing a long, pointed, fleshy nose, totally unlike that of all other monkeys. Another interesting species is the black and white entellus monkey of India, called "Hanuman" by the Hindoos, and considered sacred by them. These animals are petted and fed, and at some of the temples numbers of them come every day for

the food which the priests, as well as the people, provide for them.

The next group of Eastern monkeys are the Macaques, which are more like baboons, and often run upon the ground. They are more bold and vicious than the others. All have cheek-pouches, and though some have long tails, in others the tail is short, or reduced to a mere stump. In some few this stump is so very short that there appears to be no tail, as in the magot of North Africa and Gibraltar, and in an allied species that inhabits Japan.

American Monkeys. The monkeys which inhabit America form three very distinct groups: 1. The Sapajous, which have prehensile or grasping tails; 2. The Sagouins, which have ordinary tails, either long or short; and, 3. The Marmosets, very small creatures, with sharp claws, long tails, which are not prehensile, and a smaller number of . teeth than all other American monkeys. Each of these three groups contains several sub-groups, or genera, which often differ remarkably from each other, and from all the monkeys of the Old World.

We will begin with the howling monkeys, which are the largest found in America, and are celebrated for the loud voice of the males. Often in the great forests of the Amazon, or Orinoco, a tremendous noise is heard in the night or early morning, as if a great assemblage of wild beasts were all roaring and screaming together. The noise may be heard

for miles, and it is louder and more piercing than that of any other animals, yet it is all produced by a single male howler sitting on the branches of some lofty tree. They are enabled to make this extraordinary noise by means of an organ that is possessed by no other animal. The. lower jaw is unusually deep, and this makes room for a hollow bony vessel about the size of a large walnut, situated under the root of the tongue, and having an opening into the windpipe by which the animal can force air into it. This increases the power of its voice, acting something like the hollow case of a violin, and producing those marvelous rolling and reverberating sounds which caused the celebrated traveler Waterton to declare that they were such as might have had their origin in the infernal regions. The howlers are large and stout-bodied monkeys with bearded faces, and very strong and powerfully grasping tails. They inhabit the wildest forests; they are very shy, and are seldom taken captive, though they are less active than many other American monkeys. Next come the spider-monkeys, so called from their slender bodies and enormously long limbs and tail. In these monkeys the tail is so long, strong, and perfect, that it completely takes the place of a fifth hand. By twisting the end of it round a branch the animal can swing freely in the air with complete safety; and this gives them a wonderful power of climbing and passing from tree to tree, because the distance they can stretch is that of the tail, body, and arm added together, and these are

all unusually long. They can also swing themselves through the air for great distances, and are thus able to pass rapidly from tree to tree without ever descending to the ground, just like the gibbons in the Malayan forests. Although capable of feats of wonderful agility, the spider-monkeys are usually slow and deliberate in their motions, and have a timid, melancholy expression, very different from that of most monkeys. Their hands are very long, but have only four fingers,, being adapted for hanging on to branches rather than for getting hold of small objects. It is said that, when they have to cross a river the trees on the opposite banks of which do not approach near enough for a leap, several of them form a chain, one hanging by its tail from a lofty overhanging branch and seizing hold of the tail of the one below it, then gradually swinging themselves backward and forward till the lower one is able to seize hold of a branch on the opposite side. He then climbs up the tree, and, when sufficiently high, the first one lets go, and the swing either carries him across to a bough on the opposite side or he climbs up over his companions.

Closely allied to the last are the woolly monkeys, which have an equally well-developed prehensile tail, but better proportioned limbs, and a thick, wholly fur of a uniform gray or brownish color. They have well-formed fingers and thumbs, both on the hands and feet, and are rather deliberate in their motions, and exceedingly tame and affectionate in

captivity. They are great eaters, and are usually very fat. They are found only in the far interior of the Amazon Valley, and, having a delicate constitution, seldom live long in Europe. These monkeys are not so fond of swinging themselves about by their tails as are the spider-monkeys, and offer more opportunities of observing how completely this organ takes the place of a fifth hand. When walking about a house, or on the deck of a ship, the partially curled tail is carried in an horizontal position on the ground, and the moment it touches anything it twists round it and brings it forward, when, if eatable, it is at once appropriated; and when fastened up the animal will obtain any food that may be out of reach of its hands with the greatest facility, picking up small bits of biscuit, nuts, etc., much as an elephant does with the tip of his trunk.

We now come to a group of monkeys whose prehensile tail is of a less perfect character, since it is covered with hair to the tip, and is of no use to pick up objects. It can, however, curl round a branch, and serves to steady the animal while sitting or feeding, but is never used to hang and swing by, in the manner so common with the spider-monkeys and their allies. These are rather small-sized animals, with round heads and with moderately long tails. They are very active and intelligent, their limbs are not so long as in the preceding group, and, though they have five fingers on each hand and foot, the hands have weak and hardly opposable thumbs.

Some species of these monkeys are often carried about by itinerant organ-men, and are taught to walk erect and perform many amusing tricks. They form the genus Cebus of naturalists.

The remainder of the American monkeys have nonprehensile tails, like those of the monkeys of the Eastern hemisphere; but they consist of several distinct groups, and differ very much in appearance and habits. First we have the Sakis, which have a bushy tail and usually very long and thick hair, something like that of a bear. Sometimes the tail is very short, appearing like a rounded tuft of hair; many of the species have fine bushy whiskers, which meet under the chin, and appear as if they had been dressed and trimmed by a barber, and the head is often covered with thick, curly hair, looking like a wig. Others, again, have the face quite red, and one has the head nearly bald a most remarkable peculiarity among monkeys. This latter species was met with by Mr. Bates on the upper Amazon, and he describes the face as being of a vivid scarlet, the body clothed from neck to tail with very long, straight, and shining white hair, while the head was nearly bald, owing to the very short crop of thin, gray hairs. As a finish to their striking physiognomy, these monkeys have bushy whiskers, of a sandy color, meeting under the chin, and yellowish-gray eyes. The color of the face is so vivid that it looks as if covered with a thick coat of bright scarlet paint. These creatures are very delicate, and

have never reached Europe alive, though several of the allied forms have lived some time in our Zoological Gardens.

An allied group consists of the elegant squirrel-monkeys, with long, straight, hairy tails, and often adorned with prettily variegated colors. They are usually small animals; some have the face marked with black and white, others have curious whiskers, and their nails are rather sharp and claw-like. They have large, round heads, and their fur is more glossy and smooth than in most other American monkeys, so that they more resemble some of the smaller monkeys of Africa. These little creatures are very active, running about the trees like squirrels, and feeding largely on insects as well as on fruit.

Closely allied to these are the small group of night-monkeys, which have large eyes, and a round face surrounded by a kind of ruff of whitish fur, so as to give it an owl like appearance, whence they are sometimes called owl-faced monkeys. They are covered with soft, gray fur, like that of a rabbit, and sleep all day long, concealed in hollow trees. The face is also marked with white patches and stripes, giving it a rather carnivorous or cat-like aspect, which, perhaps, serves as a protection, by causing the defenseless creature to be taken for an arboreal tiger-cat, or some such beast of prey.

This finishes the series of such of the American monkeys as have a larger number of teeth than those of the Old World. But there is another group, the Marmosets, which have the same number of teeth as Eastern monkeys, but differently

distributed in the jaws, a premolar being substituted for a molar tooth. In other particulars they resemble the rest of the American monkeys. These are very small and delicate creatures, some having the body only seven inches long. The thumb of the hands is not opposable, and instead of nails they have sharp, compressed claws. These diminutive monkeys have long, non-prehensile tails, and they have a silky fur, often of varied and beautiful colors. Some are striped with gray and white, or are of rich brown or golden-brown tints, varied by having the head or shoulders white or black, while in many there are crests, frills, manes, or long ear-tufts, adding greatly to their variety and beauty. These little animals are timid and restless; their motions are more like those of a squirrel than a monkey. Their sharp claws enable them to run quickly along the branches, but they seldom leap from bough to bough, like the larger monkeys. They live on fruits and insects, but are much afraid of wasps, which they are said to recognize even in a picture. This completes our sketch of the American monkeys, and we see that, although they possess no such remarkable forms as the gorilla or the baboons, yet they exhibit a wonderful diversity of external characters, considering that all seem equally adapted to a purely arboreal life. In the howlers we have a specially developed voice-organ, which is altogether peculiar; in the spider-monkeys we find the adaptation to active motion among the topmost branches of the forest-

trees carried to an extreme point of development; while the singular nocturnal monkeys, the active squirrel-monkeys, and the exquisite little marmosets, show how distinct are the forms under which the same general type may be exhibited, and in how many varied ways existence may be sustained under almost identical conditions.

Lemurs. In the general term, monkeys, considered as equivalent to the order Primates, or the Quadrumana of naturalists, we have to include another sub-type, that of the lemurs. These animals are of a lower grade than the true monkeys, from which they differ in so many points of structure that they are considered to form a distinct sub-order, or, by some naturalists, even a separate order. They have usually a much larger head and more pointed muzzle than monkeys; they vary considerably in the number, form, and arrangement of the teeth; their thumbs are always well developed, but their fingers vary much in size and length; their tails are usually long, but several species have no tail whatever, and they are clothed with a more or less woolly fur, often prettily variegated with white and black. They inhabit the deep forests of Africa, Madagascar, and Southern Asia, and are more sluggish in their movements than true monkeys, most of them being of nocturnal or crepuscular habits. They feed largely on insects, eating also fruits and the eggs or young of birds.

The most curious species are the slow lemurs of South

India, small tailless nocturnal animals, somewhat resembling sloths in appearance and almost as deliberate in their movements, except when in the act of seizing their insect prey; the tarsier, or specter-lemur, of the Malay Islands, a small long-tailed nocturnal lemur, remarkable for the curious development of the hind-feet, which have two of the toes very short and with sharp claws, while the others have nails, the third toe being exceedingly long and slender, though the thumb is very large, giving the feet a very irregular and outre appearance; and, lastly, the aye-aye of Madagascar, the most remarkable of all. This animal has very large ears and a squirrel-like tail, with long, spreading hair. It has large curved incisor teeth, which add to its squirrel-like appearance and caused the early naturalists to class it among the rodents. But its most remarkable character is found in its fore-feet or hands, the fingers of which are all very long and armed with sharp, curved claws, but one of them, the second, is wonderfully slender, being not half the thickness of the others. This curious combination of characters shows that the aye-aye is a very specialized form that is, one whose organization has been slowly modified to fit it for a peculiar mode of life. From information received from its native country, and from a profound study of its organization, Professor Owen believes that it is adapted for the one purpose of feeding on small, wood-boring insects. Its large feet and sharp claws enable it to cling firmly to

the branches of trees in almost any position; by means of its large, delicate ears it listens for the sound of the insect gnawing within the branch, and is thus able to fix its exact position; with its powerful curved gnawing teeth it rapidly cuts away the bark and wood till it . exposes the burrow of the insect, most probably the soft larva of some beetle, and then comes into play the extraordinary long wire-like finger, which enters the small cylindrical burrow, and with the sharp bent claw hooks out the grub. Here we have a most complex adaptation of different parts and organs all converging to one special end, that end being the same as is reached by a group of birds, the wood-peckers, in a different way; and it is a most interesting fact that, although woodpeckers abound in all the great continents, and are especially common in the tropical forests of Asia, Africa, and America, they are quite absent from Madagascar. We may therefore consider that the aye-aye really occupies the same place in nature in the forests of this tropical island as do the woodpeckers in other parts of the world.

Distribution, Affinities, and Zoological Rank of Monkets. Having thus sketched an outline of the monkey-tribe as regards their more prominent external characters and habits, we must say a few words on their general relations as a distinct order of mammalia. No other group, so extensive and so varied as this, is so exclusively tropical in its distribution, a circumstance no doubt due to the fact

that monkeys depend so largely on fruit and insects for their subsistence. A very few species extend into the warmer parts of the temperate zones, their extreme limits in the northern hemisphere being Gibraltar, the western Himalayas at eleven thousand feet elevation, East Thibet, and Japan. In America they are found in Mexico, but do not appear to pass beyond the tropic. In the southern hemisphere they are limited by the extent of the forests in South Brazil, which reach about 30 south latitude. In the East, owing to their entire absence from Australia, they do not reach the tropic; but in Africa some baboons range to the southern extremity of the continent.

But this extreme restriction of the order to almost tropical lands is only recent. Directly we go back to the Pliocene period of geology, we find the remains of monkeys in France, and even in England. In the earlier Miocene several kinds, some of large size, lived in France, Germany, and Greece, all more or less closely allied to living forms of Asia and Africa. About the same period monkeys of the South American type inhabited the United States. In the remote Eocene period the same temperate lands were inhabited by lemurs in the East, and by curious animals believed to be intermediate between lemurs and marmosets in the West. We know from a variety of other evidence that throughout these vast periods a mild and almost sub-tropical climate extended over all Central Europe and parts of North

America, while one of a temperate character prevailed as far north as the Arctic Circle. The monkey tribe then enjoyed a far greater range over the earth, and perhaps filled a more important place in Nature than it does now. Its restriction to the comparatively narrow limits of the tropics is no doubt mainly due to the great alteration of climate which occurred at the close of the Tertiary period, but it may have been aided by the continuous development of varied forms of mammalian life better fitted for the contrasted seasons and deciduous vegetation of the north temperate regions. The more extensive area formerly inhabited by the monkey-tribe would have favored their development into a number of divergent forms, in distant regions and adapted to distinct modes of life. As these retreated southward and became concentrated in a more limited area, such as were able to maintain themselves became mingled together as we now find them, the ancient and lowly marmosets and lemurs subsisting side by side with the more recent and more highly developed howlers and anthropoid apes.

Throughout the long ages of the Tertiary period monkeys must have been very abundant and very varied, yet it is but rarely that their fossil remains are found. This, however, is not difficult to explain. The deposits in which mammalian remains most abound are those formed in lakes or in caverns. In the former the bodies of large numbers of terrestrial animals were annually deposited, owing to

their having been caught by floods in the tributary streams, swallowed up iu marginal bogs or quicksands, or drowned by the giving way of ice. Caverns were the haunts of hyenas, tigers, bears, and other beasts of prey, which dragged into them the bodies of their victims, and left many of their bones to become imbedded in stalagmite or in the muddy deposit left by floods, while herbivorous animals were often carried into them by these floods, or by falling down the swallow-holes which often open into caverns from above. But, owing to their arboreal habits, monkeys were to a great extent freed from all these dangers. Whether devoured by beasts or birds of prey, or dying a natural death, their bones would usually be left on dry land, where they would slowly decay under atmospheric influences. Only under very exceptional circumstances would they become imbedded in aqueous deposits; and, instead of being surprised at their rarity, we should rather wonder that so many have been discovered in a fossil state.

Monkeys, as a whole, form a very isolated group, having no near relations to any other mammalia. This is undoubtedly an indication of great antiquity. The peculiar type which has since reached so high a development must have branched off the great mammalian stock at a very remote epoch, certainly far back in the Secondary period, since in the Eocene we find lemurs and lemurine monkeys already special-ized. At this remoter period they were probably not separable from the

insectivora, or (perhaps) from the ancestral marsupials. Even now we have one living form, the curious Galeopithecus, or flying lemur, which has only recently been separated from the lemurs, with which it was formerly united, to be classed as one of the insectivora; and it is only among the opossums and some other marsupials that we again find hand-like feet with opposable thumbs, which are such a curious and constant feature of the monkey-tribe.

This relationship to the lowest of the mammalian tribes seems inconsistent with the place usually accorded to these animals at the head of the entire mammalian series, and opens up the question whether this is a real superiority or whether it depends merely on the obvious relationship to ourselves. If we could suppose a being gifted with high intelligence, but with a form totally unlike that of man, to have visited the earth before man existed in order to study the various forms of animal life that were found there, we can hardly think he would have placed the monkey-tribe so high as we do. He would ob-serve that their whole organization was specially adapted to an arboreal life, and this specialization would be rather against their claiming the first rank among terrestrial creatures. Neither in size, nor strength, nor beauty, would they compare with many other forms, while in intelligence they would not surpass, even if they equaled, the horse or the beaver. The carnivora, as a whole, would certainly be held to surpass them in the exquisite perfection of their

physical structure, while the flexible trunk of the elephant, combined with his vast strength and admirable sagacity, would probably gain for him the first rank in the animal creation.

But if this would have been a true estimate, the mere fact that the ape is our nearest relation does not necessarily oblige us to come to any other conclusion. Man is undoubtedly the most perfect of all animals, but he is so solely in respect of characters in which he *differs* from all the monkey-tribe—the easily erect posture, the perfect freedom of the hands from all part in locomotion, the large size and complete opposability of the thumb, and the well-developed brain, which enables him fully to utilize these combined physical advantages. The monkeys have none of these, and without them the amount of resemblance they have to us is no advantage, and confers no rank. We are biased by the too exclusive consideration of the man-like apes. If these did not exist, the remaining monkeys could not be thereby deteriorated as to their organization or lowered in their zoölogical position; but it is doubtful if we should then class them so high as we now do. We might then dwell more on their resemblances to lower types—to rodents, to insectivora, and to marsupials, and should hardly rank the hideous baboon above the graceful leopard or stately stag. The true conclusion appears to be, that the combination of external characters and internal structure which exists in the monkeys is that which, when

greatly improved, refined, and beautified, was best calculated to become the perfect instrument of the human intellect, and to aid in the development of man's higher nature; while, on the other hand, in the rude, inharmonious, and undeveloped state which it has reached in the quadrumana, it is by no means worthy of the highest place, or can be held to exhibit the most perfect development of existing animal life.—*Contemporary Review.*

www.ingramcontent.com/pod-product-compliance
Lightning Source LLC
Chambersburg PA
CBHW022343280326
41934CB00006B/760
* 9 7 8 1 4 7 3 3 2 9 6 5 2 *

www.ingramcontent.com/pod-product-compliance
Lightning Source LLC
Chambersburg PA
CBHW022343280326
41934CB00006B/758